David Hendricks Bergey

The Biological Relation Between Bacteria and the More Highly Organized Flora of Running Streams

David Hendricks Bergey

The Biological Relation Between Bacteria and the More Highly Organized Flora of Running Streams

ISBN/EAN: 9783744689939

Printed in Europe, USA, Canada, Australia, Japan

Cover: Foto ©berggeist007 / pixelio.de

More available books at **www.hansebooks.com**

Publications

OF THE

University of Pennsylvania.

New Series, No. 4.

Contributions

from

The Laboratory of Hygiene.

No. 1-2.

The Biological Relation between Bacteria and the More Highly Organized Flora of Running Streams.

D. H. BERGEY, M. D.

Comparative Studies upon the Pseudo-diphtheria, or Hofmann Bacillus, the Xerosis Bacillus, and the Löffler Bacillus.

D. H. BERGEY, M. D.

PUBLISHED FOR THE UNIVERSITY OF PENNSYLVANIA.

PHILADELPHIA.

1898.

THE
BIOLOGICAL RELATION BETWEEN BACTERIA

AND THE

MORE HIGHLY ORGANIZED FLORA OF RUNNING STREAMS

BY

D. H. BERGEY, M. D., *First Assistant*

Laboratory of Hygiene, University of Pennsylvania

THE BIOLOGICAL RELATION BETWEEN BACTERIA AND THE MORE HIGHLY ORGANIZED FLORA OF RUNNING STREAMS.

By D. H. Bergey, M. D.

First Assistant, Laboratory of Hygiene, University of Pennsylvania.

[Read before the Biological Club, March 7, 1898.]

All cryptogamic plants may be divided into two great classes with reference to the presence or absence of chlorophyl within the plant cells. When we come to study the life history of these two groups of organisms we find that their biological functions are closely related to the fact whether they contain chlorophyl or not.

In many respects the biological functions of the two groups of organisms are entirely different from each other. Those cryptogamic plants which contain chlorophyl in their cells possess biological functions not markedly dissimilar in character to those of the phanerogamic plants. Those of the cryptogamic plants which are without chlorophyl have biological functions which are entirely different from those of the phanerogamic plants.

As the basis of our study of the relation between the biological functions of bacteria and the more highly organized flora of running streams we shall take into consideration the nature of their food supply. The food supply of the more highly organized flora does not differ in any essential respect from that of the phanerogamic plants. They subsist on the various mineral matters contained in the soil and water and upon the constituents of the atmosphere. For the healthy action of all their body cells they require, in addition, the presence of sunlight. Their chlorophyl-bearing cells are dependent upon the influence of sunlight for the power of building up the most complex organic compounds and elaborating them into the various tissues and special organs. Their food supply consists of dissolved or soluble salts of various kinds in the soil and water, and of the carbonic acid of the atmosphere.

On the other hand, the bacteria and other chlorophyl-free plants subsist mainly upon nitrogenous organic matter or upon inorganic compounds containing nitrogen, such as the various salts of ammonia and the nitrates, from which they derive their supply of nitrogen. In most instances the nitrogen is most readily taken from diffusible albuminoid matter; less easily from the ammonia combinations.

Nägeli was able to demonstrate that a certain class of bacteria known as the denitrifying bacteria were able to bring about the reduction of nitrates and convert them into nitrites, ammonia, or even nitrogen.

A quite exceptional mode of meeting the demand for nitrogen is found in those bacteria forming the tubercles on the leguminoseae and allied plants in which they, in symbiosis with the host plants on which they live, are capable of fixing the free nitrogen of the atmosphere and assimilating the same in building up their cell bodies.

The supply of carbon of the chlorophyl-free plants, in contradistinction to that of the higher plants, is obtained by breaking up different forms of carbo-hydrates. Aside from albumen and peptone, they use sugar and similar carbo-hydrates and glycerine, as a source of carbon. They are also capable of using organic matters of different chemical constitution, as the basic and diabasic acids (vinegar), the hydroxilized acids (tartaric and citric), as asparagin, lucin, the different alcohols, amine, ester, urea, etc. In very dilute solutions they are even capable of using as nourishment compounds containing carbon which, in concentrated solutions, are distinctly poisonous, such as carbolic acid and salicylic acid.

Some of the chlorophyl-bearing organisms, such as the algae, which closely resemble the bacteria in some respects, ordinarily satisfy their requirements for carbon and nitrogen by taking these elements from CO_2 and NH_3, or even HNO_3, out of substances in their environment, and converting them into complex carbonaceous and nitrogenous matter with the aid of the chlorophyl. It is possible for these plants to take their nourishment out of water containing the necessary mineral substances, and out of air containing CO_2 and NH_3 and assimilate the carbon and nitrogen derived from these sources. They also possess, to a limited extent, the power of breaking up organic matter and derive their C and N directly from such matter. The bacteria, however, through the absence of chlorophyl, are incapable of existing in this manner, but require previously-prepared organic substances to meet the needs of their bodies, and to form new substance. Therefore, they cannot exist in pure water containing only mineral matter. They vegetate on dead organic matter, rich in carbon and nitrogen, and likewise on dead plants and animal organisms, or they live as parasites drawing the organic matter necessary for their life and growth from their vegetable and animal hosts.

From this one may form an idea as to the significance of the bacteria in nature. In order to provide continuously the simple nourishment necessary for the chlorophyl-bearing vegetation, it is essential to have a constant reduction and solution of this form of vegetable substances back into simple chemical combinations. The entire yearly vegetation which has formed and died must be so changed in a relatively short time that the complicated vegetable substances, the albumen, carbo-hydrates and cellulose are again converted into water, CO_2 and NH_3. Only under such conditions is the continuous regeneration of the higher vegetation possible.

A part of this work of reconstruction falls upon animal organisms, the animal cells breaking up the vegetable matter taken up, and delivering it over for oxidation.

The energy which is stored in the complex chemical combinations of the plants of higher organization with the aid of their chlorophyl and sunlight is used by the animal organisms for the production of animal heat and the different functions of the body.

With this very brief consideration of the general biological functions of the bacteria and the more highly organized flora, we are in a position to consider briefly the relation which these organisms bear to each other in running streams. This relation has been studied to some extent by experimenters with the view of discovering something about the modus operandi of the self-purification of polluted streams. It has been found that the chlorophyl-bearing organisms are confined largely to fresh, running water. These organisms do not multiply very rapidly, and may even die off, in polluted waters. The conditions for growth in waters of this character are not very favorable for these organisms. The greater the degree of pollution the more the action of sunlight—the most essential feature of the life and growth of chlorophyl-bearing plants —is interfered with in these waters. The character of the food material is not adapted to their growth because of the predominance of organic substances in the water. Besides this, because of the presence of large quantities of organic matter, we have immense numbers and different kinds of chlorophyl-free organisms in water of this character, and these are doubtless in a certain degree antagonistic to the chlorophyl-bearing organisms : so that their growth is interfered with in this manner.

As has been said, the large quantity of organic matter in polluted water, the hindrance of the action of sunlight, and the presence of suspended matters, makes this a favorable soil for the growth and multiplication of all chlorophyl-free organisms. They find here an abundant food supply and abundance of oxygen, and all other conditions favorable for growth. In the studies which have been made

on the question of the self-purification of streams examinations of polluted waters have shown that the organisms
which predominate in these waters are the chlorophyl-free
organisms. Experiments have shown that these are capable of existing upon organic matters such as found in
these polluted streams. Certain forms have been found
more constantly than others; appearing to possess a pre-
dilection for such water. For instance, Pfeiffer and Eisen-
lohr (*Archiv für Hygiene*, Bd. 14, p. 90, 1892) in their
investigation of the river Isar found large quantities of a
particular form of beggiatoa. Samples of water collected
at different points of the river were found to contain
enormous quantities of active growths of beggiatoa roseo-
persicina (Zopf). Between the threads of this organism,
however, other fission fungi were found in enormous numbers, as yeast, diatoms, protozoa—in fact, all possible forms
of lower vegetable and animal life—forms, however, which
were not examined further because they had no significance with regard to the question under consideration.
From their investigations it seemed evident that this
particular variety of beggiatoa was present only in polluted
waters, and did not penetrate into the fresh water streams
entering the river; its great prevalence at any point
appeared to indicate the presence of undue quantities of
organic matter.

Schenk (*Centralblatt für allgemeine Gesundheitspflege*,
12. Jahr., p. 365, 1892) in his investigation on the significance of the Rhine vegetation in the self-purification of
the river, reports on several forms of algae found adhering
to the stones along the banks in large quantities, which
seemed to him to have some influence upon the purification
of the river. Among the chlorophyl-free plants he says
that the saprophytic water plants of the first and greatest
importance are the bacteria. The two forms which he
says require special consideration, from the large numbers
found, are the beggiatoa alba and the cladothrix dicho-

toma. He says that the beggiatoa is of great importance with respect to the subject under consideration, occurring in especially large quantities beneath the different sewer openings, forming a slimy layer, sometimes floating in the water, and covering everything along the banks. The masses consist of innumerable single threads built up of short members which contain numerous sulphur granules in the interior of the cells. Though distributed everywhere the organism attained its greatest development only in sewers or in grossly polluted water. The cladothrix dichotoma was usually found along the banks in connection with the beggiatoa and in connection with the cladothrix glomerata, in polluted water. These fission fungi form the most important part of the chlorophyl-free water fungi along the polluted banks.

With regard to the algae he cannot coincide with the opinion of M. von Pettenkofer, that they play an important rôle in the purification of the Rhine. He states that J. Ufflemann had also reached the conviction that the rôle of the algae in the purification of rivers should not be over-estimated, because it was established that the green-thread algae and most of the diatoms can exist only in fresh or slightly polluted water, and that their action would fail where it is most desired. It is otherwise, however, with the chlorophyl-free and saprophytic bacteria. The water bacteria in their totality play the principle rôle in the purification of rivers, as far as living organisms are concerned. Among these the beggiatoa alba have especial significance.

In a brief report on the self-purification of streams, in the Report of the Massachusetts State Board of Health for 1890, p. 793, it is stated that the micro-organisms which were instrumental in bringing about a ten-fold reduction of nitrates in the water and great changes in nitrogen compounds, with almost total removal of free ammonia, were " enormous growth of anabaena, synedra and zoospores, together with many other organisms in smaller numbers. The change produced in the water by these organisms made the water turbid and green, and consequently unfit for drinking, on account of the taste, odor and appearance. The nitrate compounds are rapidly transformed by the organic growths, and it seems reasonable to conclude that this means the destruction of substances undergoing decomposition contained in the water." They state further that "from a health standpoint this may be classed as an instance of the self-purification of streams, even though the water is temporarily rendered obnoxious by the organisms which have effected the change."

Recently E. Goldschmidt, W. Prausnitz and others have published (*Hygienische Rundschau*, 8. Jahrg., p. 161) an important investigation on the death of micro-organisms in the self-purification of rivers. Their experiments were made upon the river Isar, making a continuous study of the condition of the Isar from 1889 to the present time, by themselves and other investigators. Samples of water were collected at various points of the Isar below Munich: for instance, at upper or lower Föhring, at Ismaning, at Freising, and at Landshut. In all these investigations it was found that there is a uniform diminution in those bacteria which grow on ordinary gelatine media, in regular gradation as one proceeds down the stream. The following results show fairly well the rate of diminution at the different points, and are taken from examinations made in December, 1893, showing at Föhring 24,097 organisms per cubic centimeter; at Ismaning, 15,065; at Freising, 7134, and at Landshut, 1976.

As the result of their investigation they have reached the following conclusions:

1. The self-purification of rivers, as shown in the disappearance of impurities, is not influenced by the activity of micro-organisms.

2. The disappearance of the micro-organisms which grow on gelatine media in polluted streams takes place during the day as well as night, and is not influenced by light, though sunlight seems to assist in killing off the bacteria.

3. The death of micro-organisms follows rapidly, and during the course of twenty kilometers, requiring about eight hours, 50 per cent of the germs disappeared.

Undoubtedly, the facts as presented by this thorough and practical investigation can be interpreted in several different ways. In the first place, it must be borne in mind that the nitrifying bacteria, those which are concerned most particularly in the purification of polluted waters, do not grow on the ordinary gelatine media: consequently these did not enter into the consideration of the bacteriological analyses made by these investigators. In the second place, through the action of sedimentation, some of the organic matter subsided and took its place on the bottom of the river; with the subsidence of the food material for the micro-organisms it is most evident that they also took the same course; so that in this manner a considerable proportion of the micro-organisms may have been removed. Furthermore, it is altogether probable that there was, and always is, a marked reduction of those bacteria which are capable of growing on gelatine media, and that they are destroyed through the purifying agencies at work in flowing streams.

The putrefactive decomposition of albuminous material of animal and vegetable origin is effected by a great variety of micro-organisms and gives rise to the formation of a great variety of products, some of which are volatile and characterized by their offensive odors. According to Flügge, the first change which occurs consists in the transformation of the albumens into peptone, and this may be effected by a large number of different bacteria. The special products of putrefaction vary according to the nature of the material and the conditions in which it is placed and the micro-organisms present. One or other of the bacteria concerned will take precedence, when circumstances favor its growth. Thus, the aerobic bacteria cannot grow unless putrefying material is freely exposed to atmospheric oxygen. The anaerobic bacteria require its exclusion. Some saprophytic bacteria grow at a low temperature; others take precedence when the temperature is high; some, no doubt, thrive only in the presence of products evolved by other species and are consequently associated with and dependent upon these species. Some are restrained in their growth sooner than others, by products evolved as the result of their own vital activity, or by that of associated organisms. Some grow in the presence of acids, and give rise to fermentation, which wholly prevents the development of their species.

The malodorous volatile products of putrefaction are to a considerable extent produced by the anaerobic species. For this reason these odors are more pronounced when masses of albuminous material undergo putrefaction in situations where the oxygen of the air has not access, or where it is displaced by carbon dioxide. The gases produced in the interior of a putrefying mass are mainly CH_4, H_2S and H. Many of the bacteria of putrefaction are facultative anaerobics; they are able to multiply either in the presence or absence of oxygen. The products formed by these differ, no doubt, according to whether they are or are not supplied with atmospheric oxygen. The decomposition due to aerobic bacteria is not attended with the same putrefaction odors as in the case of the anaerobic organisms; the products evolved being of a simple chemical composition,—CO_2, NH_3.

The most common organisms in running and stagnant water are the leptotricheae and the cladotricheae. The former includes the four genera: Crenothrix, Beggiatoa, Phragmidiothrix, and Leptothrix,—the latter a single

genus, Cladothrix. These organisms are widely distributed, and are found both in salt and fresh water containing
decomposing animal or vegetable material; in sulphurous
waters the beggiatoa are especially abundant and accumu-
late upon the muddy bottom, or upon organic substances
undergoing decomposition. They cover the bottom of
ponds or of small bays, forming different colored layers of
an extended and abundant growth. These organisms are
found plentifully in the refuse waters of sugar refineries
and upon the surface of putrefying vegetable and animal
material, in fresh or salt waters. The cladothrix dicho-
mata is frequently associated with the beggiatoa and is
common in the refuse water of factories, especially sugar
factories. It may readily be obtained from the surface of
putrefying algae or animal substances immersed in river
or swamp water.

As far as can be determined with the limited experimental evidence at hand, it appears evident that the first
stage of the reduction and solution of vegetable and animal
organic matters in running streams is traceable to the single
or combined action of the crenothrix, beggiatoa and cladothrix. After these organisms have in part broken up
the organic matters, with the assistance of the lower forms
of animal life also present in the water, the subsequent
stages in the operation are traceable to the action of a
species of bacteria of which the nitronomas of Winogradsky and the nitrifying bacillus of Winogradsky are,
perhaps, the best-known members. These organisms in
turn attack the products of decomposition produced by
the beggiatoa and cladothrix, and convert them into nitric
acid. These nitrifying bacteria, as they are called, are
the principal agents concerned in the reduction of organic
matter and its conversion back into inorganic matter.
They are the active agents upon which the efficiency of
sand-filtration in the purification of water supplies and of
sewage is dependent. These organisms form a slimy layer

composed of zoogloa masses of bacilli covering the surface of the filtering bed and restrain the passage of all
organic matter, both living and dead, and break it up
into inorganic salts.

The relation of these lower forms of micro-organisms,
all of which are chlorophyl-free, to the chlorophyl-bearing
organisms found in running streams is a most important
one.   Through their operation the organic impurities are
broken up and mineralized, so that in turn they can again
serve as food material for the chlorophyl-bearing organisms.
The relation of these two groups of organisms in running
streams is similar to that of these organisms when found
in the soil and elsewhere.   There is no doubt that were
all of the chlorophyl-free organisms to be removed from
the face of the earth, the life history of higher vegetation
would be a comparatively brief one; being dependent
upon the continuous supply of the mineral salts capable
of serving as food material upon the activity of the
chlorophyl-free organisms.   As soon as all of this available food material would be used up their life history
would be completed, and in turn the animal world would
be brought to the same termination as soon as all vegetable
life had ceased.

COMPARATIVE STUDIES

PSEUDO-DIPHTHERIA, OR HOFMANN BACILLUS, THE XEROSIS BACILLUS, AND THE LÖFFLER BACILLUS

D. H. BERGEY, M. D., *First Assistant*
Laboratory of Hygiene, University of Pennsylvania

COMPARATIVE STUDIES UPON THE PSEUDO-DIPHTHERIA, OR HOFMANN BACILLUS, THE XEROSIS BACILLUS, AND THE LÖFFLER BACILLUS.

By D. H. Bergey, M. D.,

First Assistant, Laboratory of Hygiene, University of Pennsylvania.

The group of pseudo-diphtheria bacilli and the xerosis bacillus are now of very great interest because, by their marked similarity to the Löffler bacillus in many of their morphological and biological characters, they lead to uncertainty in the diagnosis of true diphtheria by microscopic examination alone. The pseudo-diphtheria bacilli are sometimes found in cases of true diphtheria in connection with the Löffler bacillus, but more frequently in cases of benign throat affections, either alone or in connection with staphylococci and streptococci. The xerosis bacillus is found very frequently on the conjunctiva of healthy persons, where it apparently gives rise to no disturbances. It is also found in the nasal cavity and the throat of persons suffering from benign inflammations of the lining membrane of these cavities. Both of these groups of organisms have also been found upon the skin of various portions of the body, without at times apparently producing any appreciable symptoms; at other times they have been found in various skin diseases, in the genitourinary organs of healthy persons, as well as in various diseased conditions of these organs.

The pseudo-diphtheria bacilli were first described by Löffler (1) and by von Hofmann-Wellenhof (2), who regarded them as being non-virulent forms of the Löffler bacillus.

Roux and Yersin (3) demonstrated that the Löffler bacillus possessed great variations in virulence, sometimes being non-virulent, and they were also of the opinion that

(19)

the pseudo-diphtheria bacilli were non-virulent forms of the Löffler bacillus. Roux and Yersin isolated the pseudo-diphtheria bacillus from the mucus of the pharynx and tonsils of children, as follows: From forty-five children suffering from various affections, not diphtheritic, fifteen times; from fifty-nine healthy children, twenty-six times. They found the bacillus in five out of seven cases of measles. Inoculations into animals never produced fatal results. At times a notable edema was produced, especially so with cultures obtained from cases of measles. They state that the organisms can only be differentiated from the Löffler bacillus by inoculations into animals. Their morphological and cultural differences prove nothing.

The xerosis bacillus was discovered in large quantities by Kuschbert and Neisser (*Breslau ärtzliche Zeitschrift*, 1883, No. 4) in a condition known as xerosis conjunctivae, and their observations have been confirmed by many others.

A. Neisser (4) reports his studies on what he calls spore-formation of xerosis bacilli and other organisms. He states that when stained nearly all the bacilli show deeply-stained poles, with an unstained centre; sometimes entire separation into two short, almost square halves. The clear space in the centre is not a spore.

Ernst (5) isolated the xerosis bacillus from a twelve-year-old boy with well-marked xerosis with hemoralopie. He describes the polar granules observed in these organisms.

Abbott (6) studied the occurrence of the pseudo-diphtheria bacillus in benign throat affections, such as acute pharyngitis, follicular tonsillitis, post-nasal catarrh, simple enlargement of the tonsils, chronic pharyngitis, subacute and chronic laryngitis, and rhinitis. Out of fifty-three patients examined forty-nine presented nothing peculiar. A variety of micro-organisms were found, most commonly the pyogenic cocci. In four cases micro-organisms were found which resembled the Löffler bacillus morphologically, but were found to be non-pathogenic.

Biggs, Park and Beebe (7) reached the conclusion that from the fact that there is no means of determining by cultural methods any difference between the Löffler bacillus and the pseudo-diphtheria bacillus, and that the only difference noticeable is that of virulence, the pseudo-diphtheria bacilli are non-virulent Löffler bacilli. They found some of the cultures to produce acid in glucose broth, while others did not. The fact that some of the micro-organisms which resemble the Löffler bacillus have acid products while others do not, together with other points of difference, led them to suggest that the name pseudo-diphtheria bacillus had been applied to two distinct organisms.

Fraenkel (8) reports on the examination of a number of organisms isolated in various forms of disease of the conjunctiva, as acute and chronic conjunctivitis, trachoma, and from the healthy conjunctiva. Among the organisms isolated there were two derived from cases of xerosis conjunctivae. He is inclined to the opinion that the organisms described as xerosis bacilli by Kuschbert and Neisser (*Deutsche Medicinische Woch.*, 1884;) Fraenkel and Franke (*Archiv für Augenheilkunde*, Bd. 7, 1887); Schreiber (*Fortschritte der Medicin*, 1888, p. 650); Ernst (5), and others, were pseudo-diphtheria bacilli. He is brought to this opinion by the fact that we find in cases of true diphtheria both virulent bacilli and others whose virulence has been weakened or lost; also by the fact that this is true of a large number of other organisms, as the pneumococcus, streptococcus pyogenes, bacillus coli communis, and others. He believes that it is not unlikely that the so-called xerosis bacilli of Neisser are nothing less than true diphtheria bacilli robbed of their virulence.

There is some difference of opinion as to the xerosis bacillus occurring in normal conjunctival secretions. On the one hand Uthoft (9) states that it is frequently present in normal eyes. He treated a child with conjunctivitis crouposa for fourteen days. A culture taken from the conjunctiva was identical with the Löffler bacillus. Fourteen days after recovery the child was again examined and a culture obtained from the conjunctiva which was identical with the first, except that it was non-virulent. These cultures were examined by Fraenkel and showed the same results. On the other hand, Franke (10) examined 128 normal conjunctivae without being able to discover the bacillus.

Abbott (11) in a paper on the etiology of membranous rhinitis states his conclusions as follows:

" We are inclined to the opinion that the term ' pseudo-diphtheritic bacillus' as applied to an organism in all respects identical with the genuine diphtheritic bacillus, save for its inability to kill guinea-pigs when inoculated subcutaneously, is a misnomer, and that it would be more nearly correct to designate this organism as the attenuated or non-virulent diphtheritic bacillus, reserving the term ' pseudo-diphtheritic' for that organism or group of organisms (for there are probably several) that is enough like the diphtheria-bacillus to attract attention, but is distinguishable from it by certain morphological and cultural peculiarities aside from the question of virulence."

Howard (12) found the organism in ulcerative endocarditis, both in the valves and in other internal organs. Bernheim (13) in one case of diphtheria found also the pseudo-diphtheria bacillus, which proved to be non-pathogenic. He noticed no necrosis nor loss in body-weight in the animals inoculated, thus differentiating between the pseudo-diphtheria bacillus and diphtheria bacilli that are of low virulence. Davolos (14) found the bacillus in a case of impetigo. Deyl (15) isolated it in fifteen cases of " chalazion " formation, and considered that it was the cause of the disease. He was able to produce such formations in animals, by means of inoculation with this organism. He also found the organisms in gonorrhoea and blenorrhoea.

It has been isolated in other situations: for instance, Neisser states that he found it in cases of vaginal discharge, ulcers of the leg, etc. Kruse and Pasquale (16), in a case of liver abscess following dysentery, found a bacillus which they named bacillus clavatus, that appeared in every respect to resemble the pseudo-diphtheria bacillus. It was non-pathogenic for guinea pigs. Babes (17) found it in eight cases of trachoma and in a case of gangrene of the lung. Bass (18) found it in two cases of chronic liver disease accompanied with disturbed vision, resembling hemoralopie and xerosis conjunctivae. The xerosis bacillus was isolated from the secretions of the conjunctival sac.

Schanz (19) concludes, from the results of his investigation, that the xerosis bacillus differs from the diphtheria bacillus in lack of virulence, and that the occurrence of the xerosis bacillus in healthy eyes may lead to diphtheria in passing from the tear-ducts into the upper air passages, where in some unknown manner it acquires virulence.

Gerber and Podack (20) believe that it is not impossible for the existence of some relationship between pseudo-diphtheria bacilli and diphtheria bacilli, in cases of rhinitis fibrosa.

Peters (21) found that the differentiation between the xerosis bacillus and the pseudo- and true diphtheria bacilli cannot be made from their morphological and cultural characters. He isolated sixteen cultures: six from cases of true diphtheria, three from cases of endemic impetiginous eczema of the face, two from old cases of recurring conjuctivitis granulosa, and one each from cases of xerosis conjunctivitis crouposa, pseudo-diphtheria of the nose, and from the healthy mucous membrane of the nose.

Spronck (22) sought to differentiate between the xerosis and pseudo-diphtheria bacilli and Löffler bacilli by means of antitoxin. He states that the antitoxin prevented the reaction of true diphtheria bacilli, but that the xerosis bacilli were still capable of producing local symptoms, such as edema and inflammation, even more so than when inoculated alone. It also produced loss of appetite, drowsiness, etc., in the animal inoculated. He inoculated guinea pigs with diphtheria antitoxin, and subsequently inoculated some with the pseudo-diphtheria bacilli and others with the true diphtheria bacillus. Those inoculated with diphtheria bacilli remained well, while those inoculated with the pseudo-diphtheria bacilli showed the local reactions observed after inoculations with these organisms. He believes this to be a reliable method of differentiating between the two organisms.

C. Fraenkel (23), in the differentiation between the true and pseudo-diphtheria bacilli, also experimented with diphtheria antitoxin, free from antiseptic material, to determine whether it could serve as a means to differentiate between the Löffler bacillus and the pseudo-diphtheria bacillus. He experimented with seven cultures of the pseudo-diphtheria bacillus derived from the eyelids of healthy persons, and from several cases of conjunctivitis crouposa ; but the results were entirely negative. He then repeated the experiments of Spronck, and obtained similar results. Variable quantities of the cultures were injected into the subcutaneous tissues of guinea pigs, in this manner establishing the character of the cultures. In the second series of experiments he also used antitoxin at the same time. The results were as follows: Some of the cultures he injected into animals, in quantities of 5 to 10 c. c., produced considerable disturbance at the point of injection in the nature of swelling and infiltration, which after two days, or at the most three days, entirely disappeared. In no instance did he observe disturbances of general nature reported by Spronck—as, loss of appetite, weakness, loss of weight, or change in temperature. Even in very large doses, up to 50 c. c., the results were approximately the same. There were no fatal results; these results indicating that the antitoxin was without influence. The result was the same where the antitoxin serum was injected before, at the same time, or after the injection of the culture. He again states his belief that it is not unlikely that the so-called xerosis bacilli are nothing else than true diphtheria bacilli robbed of their virulence.

Eyre (24) sums up his investigations by stating that "in differentiating the xerosis bacillus from the Löffler bacillus we are saved all trouble in the case of first cultures by the fact that the former does not grow on blood serum at 37° C. under thirty-six to forty-eight hours, whilst the latter makes its appearance in eighteen to twenty-four hours. At the other extreme, with cultures some fifteen to twenty generations old, there is likewise very little difficulty in distinguishing between these two organisms, as the xerosis bacillus then appears as a much shorter, more slender and more curved bacillus, exhibiting neither segmentation nor clubbing. But in the case of early sub-cultures from the first culture, the circumstances are entirely altered, and we have to deal with an organism closely resembling in its general characters and mode of growth the Löffler bacillus —an organism moreover which has no one single persistent peculiarity which will enable us to say definitely, this is the xerosis bacillus. We have therefore to depend upon the sum total of the cultural and morphological differences —minute in themselves—picked out during the course of numerous observations.

As to the exact nature of the xerosis bacillus—whether it be a non-virulent and slightly altered species of the bacillus diphtheriae or a totally separate and distinct bacillus—it is impossible at present to decide.

Schanz (25) reports that in ten normal conjunctival sacs he found the xerosis bacillus four times; finding it to be the most frequent organism of the conjunctival sac. In comparing one derived from a case of xerosis with keratomalacia with the Löffler bacillus, he found that a certain differentiation of the two organisms could not be made, on the ground of their cultural and morphological characters. He states that the lack of virulence on the part of the xerosis bacillus cannot be taken as a differentiating feature, because the true diphtheria bacillus is also at times non-virulent. He believes the xerosis bacillus to be, as long as no further point of differentiation is known, a diphtheria bacillus of low virulence, and holds it to be identical with the pseudo-diphtheria bacillus.

The Massachusetts State Board of Health (26) reports on some cultures isolated in the routine examination of cultures taken from suspected cases of diphtheria. "Four of the forty-six cultures isolated were found to be pseudo-diphtheria bacilli. A few cylindrical-, pear- and hour-glass shaped bacilli are occasionally seen, but involution forms are not marked, even in old cultures. They are distinguished from diphtheria bacilli by being shorter, smaller, more uniform in size, shape and manner of staining and, as pointed out by Escherich, by a tendency to lie parallel in cover-slip preparations. These bacilli are of occasional occurrence, both in the throats of persons suffering from non-diphtheritic throat affections and in true diphtheria mingled with the Löffler bacillus. It is only in convalescent cases of long duration that the pseudo-diphtheria bacilli are likely to cause doubt. They might be mistaken for the last few remaining diphtheria bacilli, or the reverse might occur. A few remaining virulent forms may be regarded as pseudo forms. Diphtheria bacilli directly from the membrane of the throat or from cultures scarcely developed sometimes resemble quite closely the pseudo-diphtheria bacilli in morphology and staining.

"The morphological differences are reinforced by at least two biological differences of importance—the absence of any power to produce acids in bouillon containing dextrose, and the lack of pathogenic power.

"Though there are these three distinctive features of pseudo-diphtheria bacilli—characteristic morphology, absence of acid and of toxic production—it is not a simple matter to recognize them as such promptly under the microscope when taken from throat cultures, unless the observer has had considerable training. It is highly probable, therefore, that Roux and Yersin in their earlier work may have mistaken pseudo-diphtheria for true diphtheria bacilli, when they found virulent and non-virulent forms together in the throats of convalescents."

E. A. Peters (27) states that the relation of the Hofmann bacillus to the true diphtheria bacillus is clear on certain points:

(*a*) "The Hoffmann bacilli resemble the diphtheria bacilli in mode of growth, and slightly in microscopical character.

(*b*) "The cases in which they are found are liable to be mistaken for mild diphtheria. The prognosis in such cases is good.

(*c*) "There is no proof forthcoming that this bacillus is an attenuated form of the diphtheria bacillus; though the short diphtheria bacillus, when it becomes non-pathogenic, tends to resemble the Hofmann bacillus."

Schanz (28) employed as a differential diagnostic method tests of the acidity of the bouillon cultures. Earlier investigators have shown that the pseudo-diphtheria bacilli are also able, like the diphtheria bacilli, to produce acid. Neisser was able to substantiate this, yet the amount of acid with the pseudo-diphtheria bacilli after growing for forty-eight hours was from four to five times smaller than with the diphtheria bacilli. Schanz found only one exception, with a xerosis culture, which showed an equal acidity with that found in the diphtheria culture. The amount of acid formed, he says, is dependent to some extent upon the amount of the culture inoculated into a tube.

Trumpp (29) differentiated between the diphtheria and pseudo-diphtheria bacilli by means of simultaneous inoculations of diphtheria toxin. The amount of toxin inoculated was less than the minimum fatal dose, so that the additional toxin generated by the diphtheria bacilli of low virulence would be sufficient to kill the animal. Control animals inoculated with like amounts of toxin remained alive. He believes that it is not impossible to increase the virulence of acid-producing bacilli, but that non-acid producing pseudo-diphtheria bacilli cannot be made virulent. He succeeded in making a non-virulent bacillus, derived from a case of empyema following measles, virulent by inoculating it along with diphtheria toxin.

Prochaska (30) reports on sixteen cases in which the pseudo-diphtheria bacilli were found. These cases were found in making diagnoses of diphtheria, and were all cases of throat infection. Thirteen of the cultures were from cases of follicular angina. Another culture came from a case of pharyngeal and nasal diphtheria. In this case the typical Löffler bacilli were also found. When grown on blood serum they could not be differentiated from Löffler bacilli by their size, and on microscopical examination resembled the pseudo-diphtheria bacilli. Inoculation into animals gave negative results. The other two cultures were isolated from the throats of children sick with scarlet fever. In each of these cases there was a typical diphtheria membrane in the throat, yet no virulent bacilli could be found.

M. Neisser (31) gives a method of staining for differentiating between the diphtheria bacilli and pseudo-diphtheria bacilli which he considers very reliable. The solutions used are, first, one gramme powdered methylene blue (Grübler) dissolved in 20 c. c. of 96 per cent alcohol; to this is added 950 c. c. of distilled water and 50 c. c. of acetic acid; and, second, two grammes of vesuvin dissolved in a litre of boiled distilled water. Filtration, especially of the latter solution, is necessary. Cover-slips prepared in the usual manner are stained in the first solution for from one to three seconds, washed in water, and stained for from three to five seconds in vesuvin, washed in water and examined. This process will show beautiful double staining of the diphtheria bacilli, when grown on Löffler's blood serum for ten to twenty hours at 34° to 35° C., while the pseudo-diphtheria bacilli show no double staining. Neisser believes it would be best to drop the name pseudo-diphtheria, and "designate as pseudo-diphtheria bacilli only those described by von Hofmann and Löffler. The only difficulty in differentiation is with the group of xerosis bacilli and a group of rather thick, short strepto-bacilli." He reports on a careful study of twenty-two bacilli resembling the Löffler bacillus. Some of these were derived from the throat, from the nose in nasal diphtheria and from the conjunctiva. When grown on blood serum the growth of the strepto-bacilli was observed. But the similarity between the two is less marked, and there is an absence of the longer forms of the bacilli. When stained by Gram's method they appear larger, and may lead to uncertainty. The xerosis bacilli show very slight growth after six hours. When stained the bacilli appear older than the diphtheria bacilli of the same age. The growth of the xerosis bacilli after ten hours is still not very marked. The picture which they present is sometimes quite similar to that presented by the diphtheria bacilli, the form altogether so; yet not like the diphtheria bacilli of this age usually appear in such cultures. When studied with the method of double staining which he describes it is only in sixteen- to twenty-hour-old cultures that it becomes possible to make a differential diagnosis; the pseudo-diphtheria bacilli showing a negative appearance, like most of the xerosis forms. At times, however, there are single individuals that take on the characteristic staining, but it is impossible to mistake the picture presented by them for that of the diphtheria bacilli. Cultures several days old (this is especially true of many of the xerosis forms) take on the double staining.

Olmacher (32) found the pseudo-diphtheria bacilli in a case of pneumonia, accompanied with purulent meningitis.

Cobbett and Phillips (33) state that "the frequent occurrence in the mouths of healthy persons of organisms which more or less resemble the diphtheria bacillus makes it desirable that we should have some test which would enable us to distinguish them from the latter. A satisfactory test of this kind has indeed not yet been found. We know of no other way of distinguishing the non-virulent *acid-producing* bacillus but by the injection of animals. On the other hand, the cultivation of a suspected diphtheria bacillus in an alkaline glucose broth gives us a ready means of excluding the other simulator of the diphtheria bacillus. This non-virulent organism, which is characterized by its inability to produce an acid reaction in such a medium, is found in the mouths of persons suspected of having diphtheria at least as frequently as is the non-virulent acid producer. We therefore venture to recommend that this simple test be applied to any organism suspected of being the Löffler bacillus."

Fraenkel (34) investigated the value of Neisser's special method of differential staining, and says that when the conditions laid down by Neisser are strictly observed in every particular this method affords a highly satisfactory diagnostic method. He found three cultures which gave a distinct, positive result with this special staining, and also rendered the culture media acid, and in other respects resembled the true diphtheria bacilli, except that they were non-pathogenic for guinea pigs, in very large doses. Two of these cultures were derived from cases of typical diphtheria, and he is undecided whether these are non-virulent Löffler bacilli or not.

Zupink (35) in a comparative investigation of some pseudo-diphtheria bacilli found the Gram method of staining was of some value in differentiating between these organisms and the Löffler bacillus. He divides the organ-

isms into the group of pseudo-diphtheria bacilli and the
Löffler group.

De Martiui (36) experimented with two cultures of the
pseudo-diphtheria bacillus, the one rendering neutial bouillon acid, the other rendering it alkaline. The former did
not grow in fluid diphtheria antitoxin, while the latter
grew well, while both grew as well on coagulated antitoxin
as on ordinary coagulated blood serum. The Löffler bacillus did not grow well in the fluid antitoxin, but grew
well after coagulation. He believes that the acid-producing pseudo-diphtheria bacillus is a degenerated Löffler
bacillus, while the non-acid producing bacillus is a different form of organism. From his studies he is inclined to
the opinion that we are not dealing with simply modifications or varieties of the same form, but with forms of
bacilli, each having its own identity.

Spronck (37) in a lengthy dissertation touches again on
the question previously discussed by him,—on the differentiation of the true and pseudo-diphtheria bacillus. He
states that the difference in growth claimed by Martini
and Nicolas in diphtheria antitoxin was not observed by
him, wherein his results coincide with those of C. Fraenkel, though he found diphtheria bacilli that grew poorly
on the serum, but there were others that grew without
difficulty. As the only method of establishing the true
character of a doubtful culture, he again advises the
use of the inoculation method with the serum. The
pseudo-bacilli produce a swelling of the subcutaneous
tissues in large doses; even when previously injected
with serum, the true bacillus is without influence.

Fraenkel (38) says that this tedious method of differentiation is now no longer necessary since the introduction
of the method of double-staining by M. Neisser. He has
during the past few months used Neisser's method on a
large number of true and pseudo-diphtheria bacilli, and is
able to substantiate the statements of Neisser. Likewise,

it can be stated with safety that a culture which fails to give the granule formation under the proper conditions is not a culture of the true diphtheria bacillus; on the other hand, a culture which gives a positive result is most probably not pseudo-diphtheria. He is not convinced that his former opinion as to the identity of the two groups is erroneous, and that we have to deal with two distinct, if very closely related, forms of bacteria.

Muir and Ritchie (39) state that "the term xerosis bacillus has been given to an organism first observed by Kuschbert and Neisser in xerosis of the conjunctiva, and which has since been found in many other affections of the conjunctiva, and even in normal conditions. Morphologically, it is practically similar to the diphtheria bacillus, and even in cultures presents very minor differences. It is, however, non-virulent to animals and, according to Eyre, does not produce an acid reaction in neutral bouillon; in this way it can be distinguished from the diphtheria bacillus."

Glücksmann (40) states that the pseudo-diphtheria bacilli are shorter and more plump than the true diphtheria bacilli, mostly pointed at both ends, wedge-shaped; less frequently they are club-shaped. In the children's hospital he found, in cases which showed no membrane, and had no fever, that the pseudo-diphtheria bacillus was present in twenty out of thirty-nine cases examined. Large quantities of a culture of the pseudo-diphtheria bacillus were inoculated into guinea pigs without producing any effects, except at times a slight local infiltration. The pseudo-diphtheria bacillus was found not to immunize the guinea pigs against the diphtheria bacillus. An animal inoculated twenty times with large doses (15 c. c.) of a culture failed to show any immunization, as it died promptly when inoculated with diphtheria. He believes that inoculation into animals is the only mode of differentiation.

Axenfeld (41) notes certain differences between these two groups of bacilli when grown on agar, bouillon and blood serum. First, the so-called xerosis bacillus grows sparingly on the agar, very slowly, sometimes showing only after three or four days. It shows as small, very dry, closely-adherent colonies ; while the Hofmann-Löffler pseudo-diphtheria bacillus grows more rapidly, in thicker layers, moist, glistening, and easily removed. In bouillon the xerosis bacillus produces no clouding, with only small flakes at the bottom. The alkalinity is not increased. The Hofmann-Löffler bacillus grows as distinct, slimy substances at the bottom of the tube, with rapid clouding of the bouillon and increase of the alkalescence. On blood serum the xerosis bacillus grows more rapidly than on agar, but still slowly, forming dry colonies. The Hofmann-Löffler bacillus grows much more rapidly and with a more moist growth, only slightly slower than the diphtheria bacillus. He says that so far attempts have been unsuccessful to change the form of one of these growths into the other through long-continued cultivation, and it is impossible to speak of entire identity ; yet he holds it as not impossible that such identity may in time be demonstrated. He is inclined to believe, contrary to Schanz, that he was dealing with two forms of one family of organisms.

Heimersdorff (42) investigated the value of Neisser's diferential staining method for the rapid diagnosis of diphtheria of the conjunctiva. At the University eye clinic, at Breslau and Rostock, all cases of conjunctivitis crouposa are tested according to the Neisser method. He has found the test most reliable when employed with fresh cultures nine to sixteen hours old. In order to obtain a positive diagnosis he advises the simultaneous inoculation into animals. So far the diagnoses made by the Neisser method of staining were substantiated by the inoculation tests. He calls attention to the importance of having a proper blood-serum culture medium on which to grow the cultures, and advises comparing all cultures with a culture of the Löffler bacillus grown on the same culture medium, under the same conditions.

He found that at times the polar granules of the diphtheria bacillus appear later than sixteen hours, and with the xerosis bacilli earlier than twenty-four hours, and states that repeated examinations may be necessary to make a positive diagnosis.

Kruse (43) under bacillus pseudo-diphthericus (Pseudo-diphtherie, Xerose bacillus), classes bacilli which resemble the diphtheria bacillus, but are not pathogenic. Besides the studies already reported, he quotes Zarinke (*Central-blatt für Bacteriologie*, Bd. VI, s. 6), Beck (*Zeitschr. für Hygiene*, Bd. VIII), N. Klein (*Centralblatt für Bacteriologie*, Bd. VII, s. 16), Goldscheider (ref. Baumgarten's *Jahrbericht*. 92) Escherich (*Diphtherie*, Wien, 94), Pflugger (*Archiv für Ophthal.*, 37), Fick (*Micro-organismen in Konjunctivalsack*, Wiesbaden, 87), who also report on studies on these organisms, which occur in 30 per cent to 60 per cent of the mouths and noses of all persons, but more frequently still on the normal or diseased conjunctiva. They occur in great masses in xerosis conjunctiva.

Kruse places in this group also the bacillus striatus albus isolated by v. Besser from the nose of healthy persons. Likewise the bacilli found by Wilde (*Bonn Hyg. Institut*) in the secretion of a case of ozena, and the organism found by Ortman (*Berlin klin. Wochenschrift*, 89) in a diphtheritic deposit on the mucous membrane of the vaginæ of pregnant women, and those found by Rugenberg (*Bonn Hyg. Institut*) in four cases of impetigo. He states that possibly the bacillus nodosus parvus which Lustgarten isolated from the urethra belongs to this group, or is identical. Also, bacillus endocarditis griseus of Weichselbaum, and bacillus erythematis uroglini of Demme.

Schültz (44) investigated the mixed infections in pulmonary tuberculosis, and found in fifteen out of thirty cases examined bacilli resembling the diphtheria bacillus. On examining these cultures when grown on the different culture media, he found an almost complete identity with the Löffler bacillus, though nearly every culture had, on one or the other culture medium, a slight characteristic, either in the form or size of the individuals, or of the culture in general. Some of the cultures were found to be pathogenic for guinea pigs. The lesions found after death were injections around point of inoculation, exudate into the internal body cavities, and hemorrhagic condition of the adrenals. Other animals died after longer intervals without any lesions indicative of diphtheria, only cachexia with slight peritonitis. The remaining cultures were found to be non-pathogenic.

Ehret (45) reports on five cases of diabetic tuberculosis in which he also found the pseudo-diphtheria bacillus.

Auckenthaler (46) investigated the value of Neisser's method of staining in the diagnosis of diphtheria. He found that whenever Löffler bacilli were present in the cultures it was always possible to demonstrate the polar granules by Neisser's method. He believes it would be better to allow ten to 'fifteen seconds for the operation of the methylene-blue stain, in preference to one to three seconds as recommended by Neisser. He states, however, that there are here and there cultures of Löffler bacilli in which the polar granules did not appear, and in such instances he advises the examination of a number of coverslip preparations. He found also isolated bacilli in pseudo-diphtheria-bacillus cultures which contained the polar granules stained by Neisser's method. He advises in all cases of doubt that cultures be made to determine the production of acid or alkali in bouillon, preferably litmus bouillon, and inoculation into animals, before a definite diagnosis is made.

During the past two years I have been engaged upon a comparative study of the pseudo-diphtheria bacillus, the xerosis bacillus, and the Löffler bacillus, with special reference to the differentiation of these groups of organisms. My observations have been made on cultures derived from various sources, as the urine in health and disease, the normal conjunctiva, the nose and throat in catarrhal inflammations, the skin in impetigo, and from secretions of the vagina in metritis.

In January, 1896, a slight polyuria with symptoms of mild cystitis, accompanied by a slight rise in temperature in the evening, led me to undertake the bacteriological study of my own urine. I at once discovered a bacillus which, in its morphological characters, closely resembled the diphtheria bacillus. Repeated examinations of the urine disclosed the same organism persisting for some time. When studied on the different culture media it was found to grow on blood serum in the form of a small, round, whitish colony, with no tendency to coalesce after many days. On agar-agar the growth was in the form of very small, pearly-white, dry, round colonies without coalescence. In gelatin stab cultures there was at times a very slight growth, as shown by the whitish line along the inoculation stab. Sometimes no growth could be observed on gelatin. On potato only a few bacilli could be found on scraping the surface after forty-eight hours—and apparently there was no growth. In bouillon there was only a very slight growth, as only a few bacilli could be found. There was no clouding of the bouillon. In litmus-milk there was only a very slight growth. At times the color of the milk was changed to a deeper blue. In rosolic acid solution there was no growth. No bacilli could be found. The color of the solution remained the same. In peptone solution there was no growth ; no bacilli could be found. There was no indol production. In the course of a month the polyuria and fever gradually

subsided, and no further attention was paid to the matter.
During May, 1897, there was a return of the symptoms,
but in milder form. On examination the same organism
was again found in the urine. It was found repeatedly for
over a month. Whether the organism really came from
the bladder or not it is impossible to state. It was found
in cultures made from the meatus, but, judging from the
symptoms, I am inclined to believe it was present in the
bladder itself. Careful study of the organisms isolated
from my urine in May revealed no biological or morphological differences from the organisms isolated in January.
Growth always took place most readily at the body temperature ; though after growth had once become established
on ordinary media it was found possible to grow the organism at the room temperature. Morphologically, when
grown on agar-agar or blood serum the organism appeared
as a short, slender rod, sometimes showing a tendency to
clubbing at the ends, which stained more deeply at the
ends than elsewhere ; usually, when stained with dilute
methylene blue, the bacilli were marked with alternate
bands of deeply-stained and unstained areas. Usually, the
longer and more club-shaped organisms showed the striæ
best. It was never seen to grow visibly on potato. In
litmus-milk there was sometimes a change to a deeper blue
color. There was not not always a growth on gelatin.
The color of the rosolic acid solution always remained un-
changed. In bouillon there was at times a very slight
flocculent sediment at the bottom of the tubes. On blood
serum no coalescence of the small round colonies was observed at any time. The bacilli appeared to stain best
with aqueous methylene blue, but also stained well with
Löffler's alkaline methylene blue. They stained moder-
ately well with aqueous fuchsin ; and more deeply with
carbol-fuchsin. They stained deeply by Gram's method.
By staining lightly with aqueous methylene blue and
counterstaining with Bismarck brown, small bluish gran-

ules were noticeable in the ends of the club-shaped bacilli, while the body of the bacillus had a faint brownish tint.

DERIVATION OF CULTURES OF THE XEROSIS BACILLUS AND OF THE PSEUDO-DIPHTHERIA BACILLUS.

Cultures 1, 2 and 3 were isolated from my own urine. Culture 4 was isolated from the urine of Mr. A., a healthy person. Cultures 5, 6 and 8 were isolated from the conjunctival sac of my own eyes. Culture 9 was isolated from the conjunctival sac of Mr. F., a healthy person; culture 10 from the conjunctival sac of Mr. H., a] healthy person. Culture 11 was isolated from a vaginal discharge taken from the case of Miss C., in the University Hospital. Culture 12 was isolated from the urine of a patient of Dr. A. Culture 13 was isolated from the urine of a case of catarrhal jaundice. These cultures were obtained during May and June, 1897. Culture 14 was isolated from the urine of Dr. P., a healthy person. Cultures 15 and 16 were isolated from the conjunctival sac of my left and right eyes respectively. Culture 17 was obtained from my own urine. Culture 18 was derived from the same source, and culture 19 from my nose, which appeared to be in a healthy condition at the time, during October and November, 1897. Culture 20 was isolated from the vaginal secretions, metritis, of an operative case in the Maternity Department of the University Hospital, that at the time was suffering from septicaemia. Streptococci were also present in large quantities. Culture 21 was derived from a small abrasion on my knee, which at the time showed a slight tendency to suppuration. Staphylococci were also present at the time. Cultures 22 and 23 were obtained from small patches of a scaly eruption (impetigo?) on my forearm which had been noticed for several months. It was accompanied with but very slight itching and was removed after several days' treatment of a 1-500 solution of bichloride of mercury, leaving a slightly brownish, glistening spot which disappeared after several days. These cultures were obtained during November, 1897.

Culture 24 was obtained December 28, 1897, from secretion of my nose, during an attack of coryza accompanied with more or less fever, lassitude, headache, and vague pains throughout the system. The organism appeared to be in pure culture, and was found plentifully on making cover-slip preparations from the secretions. Culture 25 was derived from the same source, a few days later.

Cultures 26 and 27 were isolated January 4, 1898, from secretions of my nose during an attack of la grippe. There was a profuse muco-purulent discharge from the nostrils, with acute pharyngitis and tonsilitis, some disphagia, considerable headache, chilliness, aching of the back and limbs, with considerable fever. The prostration was greater than in the attack just preceding this. Along with these organisms were found streptococci in considerable abundance. The nature of the symptoms and the character of the discharge were markedly different from those of the previous attack. Whether these differences were due to the presence of the streptococcus in addition to the pseudo-diphtheria bacillus, or whether they were due to the streptococcus alone, or to greater virulence of the pseudo-diphtheria bacillus, it is impossible to say.

Cultures 28 and 29 were again derived from the eruption on my arm, January 31, 1898, and culture 30 from my nose, February 14, at the beginning of another attack of la grippe. The influenza bacillus was also present in large numbers, and during the attack the xerosis bacillus disappeared from the secretions of the nose. Culture 31 was derived from the urine of Mr. H., a case of mild cystitis, April 7, 1898 ; staphylococci were also present.

Morphologically, these different organisms resemble each other very closely, with the exception of culture 13, which, when first isolated, showed very large club-shaped bacilli in blood-serum cultures, but after growing on artificial media for some months it showed only as short, plump bacilli with rounded ends, the centre of which did not take on stain, while the poles were deeply stained.

All the other cultures occur as short bacilli with pointed ends, sometimes having only one clear, unstained space in the centre and the poles deeply stained; again at times they are made up of alternate bands of stained and unstained material. In blood-serum cultures and in alkali-peptone media a considerable number of club-shaped bacilli were to be seen, especially in cultures 20, 26, 27 and 29; less markedly so in cultures 18, 21, 23, 24, 30 and 31; while the remaining cultures only showed club-shaped bacilli occasionally. A peculiarity of these cultures which had been noted by other investigators, is the tendency they have of two or three lying side by side in cover-slip preparations.

Biologically, these cultures show greater differences. A most striking difference is noted in the growth on agar-agar. Cultures 20, 26, 27 and 29 grow as a thick, moist, glistening, white or yellowish-white layer on the surface of the agar. Culture 13 grows as a thick, moist, glistening, yellowish layer on agar-agar. All the other cultures grow as very minute, dry, pearly-white colonies and show no tendency to coalesce, except when sown very thickly, when they grow as a dull, pearly-white layer. These differences in growth were somewhat less marked on blood-serum, though they all manifested very much the same characteristics as on agar-agar.

In alkali-peptone bouillon (47) the same cultures show characteristics which correspond with the characteristics on agar-agar and on blood-serum. Cultures 20, 26, 27 and 29 grow with a thick yellowish-white mycoderm, which sinks to the bottom of the tube after several days, while cultures 18 and 21 show a very thin mycoderm. The remaining cultures show a very fine flocculent deposit along the sides and at the bottom of the tube, but no mycoderm.

When grown in litmus-milk considerable variation was noticed with regard to the reaction of the milk after growing at 34° C. for ten days. Under the same conditions cultures 13, 14, 19, 23, 25, 26, 27 and 29 produced an alkaline reaction which was noticeable in most instances after twenty-four hours, while cultures 5, 15, 16, 18, 20, 21 and 24 produced an acid reaction which was noticeable only after growing for four days.

When grown in neutral litmus glucose bouillon more satisfactory results were obtained with regard to the production of acids, by the different cultures, than with the use of litmus milk. With the exception of cultures 23, 26, 27 and 29, all produced an acid reaction in this medium, after growing in the incubator for ninety-six hours. Some of the cultures showed a slight acid reaction after seventy-two hours, while culture 20 showed a marked acid reaction after twenty-four hours' growth. The extent of acid production by this culture was apparently equal to that of two cultures of the Löffler bacillus grown under the same conditions. Cultures 21 and 31 produced almost the same degree of acidity as the cultures of the Löffler bacillus, though not as rapidly as in the case of culture 20. Cultures 23 and 29 failed to change the reaction of the medium. Cultures 26 and 27 showed a distinct alkaline reaction after growing ninety-six hours. These were the only cultures which produced an alkaline reaction in this medium.

When grown on potato, cultures 5, 18, 20, 21, 23, 24, 26 and 29 showed a slight growth after ten days at 34° C. A similar growth was noticed in cultures of the Löffler bacillus when grown under the same conditions. With the exception of cultures 20 and 21 the growth was barely perceptible. It is probable that the potato was somewhat alkaline, as the Löffler bacillus does not grow on acid potato.

The colonies on agar plates of cultures 13, 20, 26, 27 and 29 were somewhat more dense than those of the Löffler bacillus, and they were also slightly yellowish in color; otherwise they resembled those of the Löffler bacillus. The colonies of the other cultures resembled those of the Löffler bacillus quite closely.

The cultures were studied for some time with regard to the effect of Neisser's differential stain. They were grown on Löffler's blood serum at a temperature of 34° C. for from ten to twenty hours, and were then examined. Cultures 5, 18, 20, 21, 23, 26 and 29 showed more than the usual number of bacilli with deeply-stained polar granules ; all the other cultures, except 13, showed only an occasional bacillus with the polar granules stained. Culture 13 seemed to show the granules stained as uniformly as the cultures of the Löffler bacillus, though the morphological characters of this bacillus are different from those of the Löffler bacillus. These results coincide with the statements of Neisser that the cultures of the pseudo-diphtheria bacillus show only an occasional bacillus with the polar granules stained. Control cultures of the Löffler bacillus grown under the same conditions always showed most of the bacilli with the polar granules stained, or even with three, four or more granules in each bacillus.

Experiments were made with a large number of different aniline stains, in watery solutions, with and without decolorizing agents, with the hope of discovering additional methods of differentiation between these groups of organisms. No very satisfactory results were obtained, though the following stains served to differentiate between the pseudo-diphtheria bacilli and the Löffler bacilli ; but the differentiation was in no case more definite than by Neisser's method :

    1. Löffler's alkaline methylene blue, one minute, wash in water.

    Lugol's solution, one minute, wash in water.

Löffler bacilli—granules blue, bacilli brown.

Pseudo-diphtheria bacilli—granules blue (if any are present), bacilli brown.

2. Koch-Ehrlich gentian violet, one minute, wash in water; 10 per cent acetic acid, one minute, wash in water.

Löffler bacilli—granules violet, bacilli decolorized.

Pseudo-diphtheria bacilli—granules violet (if any are present), bacilli decolorized.

3. Watery solution of dahlia, one minute.

Löffler bacilli—granules deeply stained, bacilli light violet.

Pseudo-diphtheria bacilli—bacilli deeply stained, clear space in centre.

4. Watery solution of methyl violet, one minute.

Löffler bacilli—granules deep violet, bacilli less deeply.

Pseudo-diphtheria bacilli—bacilli deep violet, homogeneous.

5. Watery solution of crystal violet, one minute.

Same results as with methyl violet.

6. Watery solution of crystal violet, one minute, wash in water; 10 per cent solution of iodin in water, one minute, wash in water.

Same results as with watery solution alone, except that the pseudo-diphtheria bacilli show the segmentation.

From the results obtained in the study of the biological characters of the different cultures it is evident that we are dealing with two distinct groups of organisms ; cultures 20, 26, 27, 29 and 31, from their growth on agar, blood serum and alkali-peptone bouillon, belong to the group of so-called pseudo-diphtheria bacilli, while all the other cultures belong to the group of xerosis bacilli, except culture 13, which I have not classified definitely, though it probably also belongs to the former group.   These groups of organisms are evidently closely related to the Löffler bacillus. I am inclined to believe, with Fraenkel, Escherich, Spronck and Trumpp, that these organisms are distinct organisms, and that they are not a virulent Löffler bacilli.   There is so far no evidence to believe that they are attenuated Löffler bacilli.   All the experimental evidence points to the opposite opinion.   Roux and Yersin, Löffler and von Hofmann-Wellenhof were inclined to believe they were non-virulent Löffler bacilli ; this opinion being not the one now generally held by bacteriologists.

TABLE I.

The following table shows the results obtained with some of the cultures which were carefully studied in their behavior on the different culture media, presenting the comparative results in a more graphic form :

Number of Culture.	Agar Colonies.	Agar Slants.	Blood Serum.	Growth on Litmus Milk.	Growth on Potato.	Growth on Alkali-peptone Bouillon.	Morphology.	Neisser's diff. Stain.	Source of Culture.
5	Like diphtheria.	Grayish white pearly colonies	Dry whitish layer.	−	+	Flocculent sediment.	.	+	Healthy conjunctiva.
12	Like diphtheria.	Grayish white pearly colonies	Dry whitish layer.	None.	−	Flocculent sediment.	.	−	Urine.
13	?	Yellowish moist layer.	Yellowish moist layer.	+	−	Flocculent sediment.	.	+	Urine. Case of cat. jaundice.
15	Like diphtheria.	Like 5.	Like 5.	−	−	Flocculent sediment.	.	−	Healthy conjunctiva.
16	Like diphtheria.	Like 5.	Like 5.	−	−	Flocculent sediment.	.	−	Healthy conjunctiva.
18	Like diphtheria.	Like 5.	Like 5.	−	+	Thin mycoderm.	.	+	Healthy conjunctiva.
19	Like diphtheria.	Like 5.	Like 5.	+	−	Like 5.	.	−	Nose.
20	More dense.	Yellowish moist layer.	Yellowish moist layer.	+	+	Thick mycoderm.	+	+	Vaginal discharge—metritis.
21	Like diphtheria.	Like 5.	Like 5.	−	+	Like 18.	.	+	Abrasion of knee.
23	Like diphtheria.	Like 5.	Like 5.	+	+	Like 5.	.	+	Eruption on arm.
24	Like diphtheria.	Like 5.	Like 5.	−	+	Like 5.	.	+	Nose, catarrh.
25	Like diphtheria.	Like 5.	Like 5.	+	−	Like 5.	.	−	Nose, catarrh.
26	More dense.	Like 20.	Like 20.	=	−	Like 20.	.	−	Nose, coryza.
27	More dense.	Like 20.	Like 20.	−	−	Like 20.	+	+	Nose, coryza.
29	More dense.	Like 20.	Like 20.	None.	+	Like 5.	.	+	Eruption, arm.

Most of the cultures under observation were cultivated in alkali-peptone bouillon (45) for several months, each culture being transplanted into fresh media at intervals of forty-eight hours, with the hope of increasing their virulence. A number of the cultures produced a slight subcutaneous edema at the point of inoculation, when first isolated, but no greater degree of virulence could be secured by the cultivation in alkali-peptone bouillon, though with most of the cultures a constant reduction in the body weight of the animal was noticed during the first week following the inoculations. The animals usually regained their weight during the second week. Occasionally an animal succumbed after inoculation, but as far as could be determined death was always due to some intercurrent affection, usually a low-grade pneumonia due to staphylococcus infection. When inoculated into the peritoneal cavity with large doses (4 to 5 c. c.) of alkali-peptone, bouillon cultures, forty-eight hours old, several of the animals died. In the autopsies on these animals slight subcutaneous edema was sometimes found. The adrenals were at times found to be slightly hyperaemic. When inoculated intra-peritoneally, there was at times considerable sanguinous fluid in the peritoneal cavity, which contained masses of the bacilli. Numerous leucocytes were found filled with the bacilli. The organisms were recovered in cultures made from the peritoneal fluid, while cultures from the different organs remained sterile. There were nearly always extensive adhesions of loops of intestine to the abdominal wall or other organs in the animals inoculated intra-peritoneally. Aside from foci of inflammation in the lungs in some cases, and occasionally congestion of the spleen, the internal organs were found normal. No glandular enlargements were noticed.

Attempts were also made to increase the virulence of the organisms by passing the cultures through several animals, but no definite results were obtained.

An attempt was made to render a number of animals immune to the several cultures by inoculating them at intervals of four to ten days with cultures of the same organism, and then after such treatment for six weeks to two months they were inoculated with cultures of the Löffler bacillus. The results obtained do not indicate that any degree of immunity had been acquired from any of the cultures employed. When fatal doses of the Löffler bacillus were inoculated subsequently, they died from experimental diphtheria.

The following are the results of the autopsies on the animals which died in consequence of inoculations of the different cultures, in the attempt to increase the virulence by passing them through successive cultures of alkali-peptone bouillon, along with the results of sections of the organs of these animals :

Autopsy on guinea pig No. 7.—Died nine and a half days after inoculation with 1 c. c. of culture 13. Not greatly emaciated ; inguinal and axillary glands appear slightly enlarged. Abdominal walls at site of inoculation are slightly hyperaemic. Liver seems somewhat hyperaemic. Spleen is enlarged and dark. Kidneys are slightly congested. Adrenals are large, pale and brittle. The lungs are congested and show large areas here and there that are completely hepatized. Cultures made from the hepatized lung show the micrococcus tetragenous. Cultures made from the other organs remained sterile. On section the liver, kidneys and spleen showed more or less congestion, aside from which nothing abnormal was noted. The lung showed extensive infiltration into the air cells, with some proliferation of the connective-tissue cells, and marked congestion.

Guinea pig No. 55.—Died twenty-two hours after receiving a second inoculation of culture 27 (4 c. c.) into the peritoneal cavity. The subcutaneous lymphatics were somewhat congested. There was a large amount of sero-sanguinous fluid in the peritoneal cavity. The leucocytes in this fluid contained large numbers of bacilli. The liver, kidneys, lungs and adrenals were also somewhat congested. The pericardial sac contained a considerable amount of clear fluid. The bacilli were found in cultures made from the peritoneal cavity. Cultures from all the organs remained sterile. On section the lung shows here and there small areas of infiltration. The adrenals show slight areas of cell necrosis and portions of the organ seem to be completely choked with blood, showing marked congestion. Aside from the congestion nothing abnormal was noted in the kidneys, liver or spleen.

Guinea pig No. 60.—Died twenty-two days after subcutaneous inoculation with 3 c. c. of an alkali-peptone bouillon culture of 25. The animal is emaciated ; the subcutaneous lymphatics somewhat hyperaemic ; the liver dark and enlarged ; the gall-bladder well filled with pale bile ; the kidneys and lungs somewhat congested. The spleen and adrenals appear normal. Cultures made from the organs remain sterile. On section the lung shows considerable cellular infiltration. The kidney, liver and spleen are congested to a considerable extent.

Guinea pig No. 68.—Died sixteen hours after a second inoculation of 3 c. c. into the peritoneal cavity of culture 20. Shows subcutaneous edema quite marked ; some fluid in the pericardial sac ; adrenals distinctly hyperaemic ; kidneys and liver congested. The retro-peritoneal glands are not enlarged. Cover-slips prepared from the peritoneal fluid show numerous clumps of the bacilli and many leucocytes loaded with the bacilli. Cover-slips from the liver, kidney and blood were negative. Cultures from the peritoneal fluid, liver and blood showed the bacilli. The liver shows here and there rather large areas of cell necrosis. In the centre of these necrotic spots there is complete disintegration of the cellular elements. Along the margins some of the cells are partially disintegrated, showing a pale nucleus, or a fragmented nucleus ; others show fairly well in outline. The lung shows here and there small areas of cellular infiltration. Aside from the congestion, nothing abnormal was noted in the kidneys.

Guinea pig No. 65.—Died ten days after being inoculated with 4 milligrams of a 24-hour-old blood-serum culture of 20. The liver and kidneys showed considerable congestion. The bladder was very much distended with pale, straw-colored urine. In the interior of the bladder was what appeared to be an old pseudo-membranous cast of the lining membrane. It was whitish in color, soft, friable, and stained imperfectly, the stained matter looking like cell debris. A few bacilli were found in this matter. The mucous membrane of the bladder was hyperaemic. Numerous bacilli were found in cover-slip preparations made from the seat of inoculation. Cultures made from the internal wall of the bladder and from the pseudo-membrane in the bladder showed the pseudo-diphtheria bacillus.

Guinea pig No. 76.—Had been inoculated at intervals for about two months, five in number, with alkali-peptone bouillon cultures of 26, with the idea of rendering it immune. The animal died about thirty hours after the last inoculation into the peritoneal cavity. On autopsy a small hemorrhagic point was observed in the abdominal wall, at the point of puncture with the needle, in which the bacilli were found. There was a large amount of sanguinous fluid in the peritoneal cavity. Adhesions of the abdominal organs and recent peritonitis were also noted. The organs were congested ; the adrenals were slightly hyperaemic. The pericardial sac also contained a little clear fluid. Cover-slips made from the peritoneal fluid showed leucocytes filled with bacilli. Cultures from the exudate in the peritoneal cavity showed the pseudo-diphtheria bacillus.

Guinea pig No. 98.—Was inoculated subcutaneously with 4 c. c. of a 48-hour-old alkali-peptone bouillon culture of 27, and died eighteen days later. On autopsy no lesion was found at the site of inoculation. There was no fluid in the internal cavities. Spleen, kidneys and liver were normal ; adrenals slightly hyperaemic. The lungs showed areas of inflammation. Cover-slips made from the lungs showed micrococci. On section, the liver, kidneys, spleen and adrenals were found normal. The lung was found to be very much conjested, with infiltration of cellular elements, showing areas of complete consolidation.

It will be observed that the results obtained in the autopsies on these animals are somewhat variable. On the whole, the lesions observed were dissimilar from those found in experimental diphtheria. In several of the animals some slight subcutaneous edema was noted, but nothing very marked. There was in no instance distinct glandular enlargement, either of the external lymphatic glands or of the retro-peritoneal glands, as noted usually in inoculations with the Löffler bacillus. The appearance of the adrenals, while at times somewhat hyperaemic, was usually the contrary— that is, they were rather pale and anaemic.

Sections of the organs of these animals presented no distinctive features. Usually, death supervened after two or more inoculations with the same organism, at intervals of five to seven days. The most common feature in these autopsies was the apparent lowered vitality of the animal, as the result of the inoculations ; and in several instances the animal evidently succumbed to secondary inflammation, more particularly inflammation of the lung, as the result of the invasion of pathogenic cocci.

The results of the inoculation experiments afford no assistance in classifying these organisms. Neither do the results of the growth in the different culture media present features which are uniformly constant with regard to the different cultures. However, from their biological characters, as noted more particularly on agar-agar and blood serum with regard to the macroscopic appearance of the cultures, we may classify them into two groups, more or less distinct, namely : those which grow as a distinct layer upon these media, forming a rather thick, creamy-white surface-growth, and the group comprising those which grow only, or usually at least, as small, pearly-white, isolated colonies on these media. The former group comprises the organism which is usually designated as the pseudo-diphtheria bacillus, or Hofmann bacillus. The latter group comprises what is usually known as the xerosis bacillus. Morphologically, these two groups of organisms resemble each other very closely, and they also resemble the Löffler bacillus quite closely,—so much so that their occurrence in cultures made for the diagnosis of diphtheria is sure to lead to errors in diagnosis unless the precaution is taken to apply Neisser's differential stain. Our experience has been, however, that where this differential staining method has been systematically employed a safe differential diagnosis can be made, from the fact that neither of these groups of organisms shows the polar granules of Neisser when grown under the prescribed conditions.

There has been for some time considerable controversy between bacteriologists with regard to the identity of these micro-organisms. Loffler, v. Hofmann, Roux and Yersin, Koch, Dunbar and Abbott are of the opinion that these two groups of organisms are merely modified forms of the Löffler bacillus. On the other hand, Hueppe, Fraenkel, Escherich, Spronck and Trumpp, among others, belonging to what is called the separatist school, hold that these two groups of organisms are entirely different from the Löffler bacillus, and that diphtheria is always produced by virulent diphtheria bacilli, and that these two groups of organisms never increase in virulence to such an extent as to produce diphtheria. Most of the later investigators are ready to accept the teaching of the separatist school, by the light of the more definite methods now being employed.

From the results of my studies of these organisms I have reached the conclusion that we have to deal with a large group of micro-organisms, at the head of which is the virulent Löffler bacillus, which may occur with several distinct variations, as shown in cultures derived from different cases of diphtheria. The most marked differences with regard to the Löffler bacillus are, first, long, slender bacilli which show a tendency to the formation of rather large, club-shaped organisms on blood-serum ; and second, rather short, thick, ovoid bacilli which show the club-shaped forms only occasionally.

At the other extreme of this large group of micro-organisms is the xerosis bacillus. Between these two extremes, the types of which are the Löffler bacillus on the one hand and the xerosis bacillus on the other, we have many variations in type, as shown by the modifications of biological and morphological characters. The organisms which I have been studying, belonging more distinctly to the group of pseudo-diphtheria, or Hoffmann bacilli, are cultures Nos. 13, 20, 26, 27, 29 and 31. All the other cultures studied belong to the group of xerosis bacilli.

I believe it would be preferable to designate as Hofmann bacilli all those so-called pseudo-diphtheria bacilli which grow as a thick, creamy-white layer on agar-agar and on blood serum. It is evident that these organisms are not a-virulent Löffler bacilli, and therefore have no direct connection with the disease processes in cases of diphtheria. If they are capable of producing any lesions whatever these are of a mild character and largely local in their manifestation. The conditions existing in my nose and throat at the time of isolating several of the cultures studied were of the nature of a mild catarrhal inflammation.

In this connection I wish to thank Dr. A. C. Abbott, Director of the Laboratory, for valuable suggestions and advice given during the investigation.

BIBLIOGRAPHY.

1. Löffler: Centralblatt für Bacteriologie. Bd. II, p. 105.
2. von Hofmann-Wellenhof: Wien. medicinische Wochenschrift. 1888. Nos. 3 and 4.
3. Roux and Yersin: Annales de l'Institut Pasteur. 1890.
4. A. Neisser: Zeitschr. für Hygiene. Bd. IV, 165.
5. Ernst: Zeitschr. für Hygiene. Bd. IV, p. 25.
6. Abbott: Johns Hopkins Hosp. Bul. 1891. Vol. II, p. 160.
7. Biggs, Park and Beebe: Diagnosis of Diphtheria in New York, 1893-94. Abst. Centralblatt für Bacteriologie. Bd. XVII, s. 765.
8. C. Fraenkel: Berliner klinische Wochenschrift. 1893. No. 11. Centralblatt für Bacteriologie. Bd. XIV, p. 364.
9. Uthoft, quoted by Schanz: Berliner klinische Wochenschrift. No. 12. 1896.
10. E. Fraenkel: Archiv. f. Ophthal. Bd. XXXIX, s. 529. 1893.
11. Abbott: The Medical News. May 13, 1893.
12. Howard: Johns Hopkins Hosp. Bul. 1893. No. 30.
13. Bernheim: Zeitschr. für Hygiene. Bd. XVIII.
14. Davolos: Cronica médico-quirirgica de la Habana. 1894. Nos. 12 and 13. Ref. Centralblatt für Bacteriologie. Bd. XVII, p. 1.
15. Deyl: Ref. Centralblatt für Bacteriologie. Bd. XIV, p. 404.
16. Kruse and Pasquale: Zeitschr. für Hygiene. Bd. XVI, p. 1.
17. Babes: Semaine médicale. 1895. p. 63.
18. Bass: von Graeffe's Arch. für Ophthalmologie. Bd. XLVIII, p. 608, Abt. 5. Centralblatt für Bacteriologie. 1895. Bd. XVII, p. 840.
19. Schanz: Deutsche medicinische Wochenschrift. 1894. No. 49. Centralblatt für Bacteriologie. 1895. Bd. XVII, p. 260.
20. Gerber and Podack: Deutsches Arch. für klinische Medicin. No. 54, p. 262. Centralblatt für Bacteriologie. 1895. Bd. XVII, p. 723.
21. Peters: Sitzungsberichte der Niederrhein. Gesellschaft für Natur- und Heilkunde zu Bonn. 1896. Centralblatt für Bacteriologie. 1896. Bd. XX, p. 595.
22. Spronck: Deutsche medicinische Wochenschrift. No. 36. Semaine médicale. 1896. No. 40. Centralblatt für Bacteriologie. 1896. Bd. XX, pp. 626 and 776.
23. C. Fraenkel: Hygienische Rundschau. Bd. VI, No. 20. 1896.
24. Eyre: The Lancet. December 21, 1895. Centralblatt für Bacteriologie. 1896. Bd. XIX, p. 729. Journal of Path. and Bacteriol. Vol. IV, p. 54. 1896.
25. Schanz: Berliner klinische Wochenschrift. 1896. No. 12. Centralblatt für Bacteriologie. Bd. XX, p. 595.

54   *Bibliography.*

26. Report of Massachusetts State Board of Health. 1896. P. 664.
27. E. A. Peters: Journal of Path. and Bacteriol. Vol. IV, p. 181.
    1896.
28. Schanz: Berliner klinische Wochenschrift. No. 50. 1897.
29. Trumpp: Centralblatt für Bacteriologie. Bd. XX, p. 721. 1896.
30. Prochaska: Zeitschr. für Hygiene. 1897. P. 373.
31. M. Neisser: Zeitschr. für Hygiene. Bd. XXIV, s. 443. 1897.
32. Olmacher: Ref. Hygienische Rundschau. 1896. Jour. American
    Med. Assoc. March 2, 1895. New York Med. Jour. April 27,
    1895.
33. Cobbett and Phillips: Journal of Path. and [Bacteriol. Vol. IV,
    p. 193.
34. C. Fraenkel: Berliner klinische Wochenschrift. No. 50. 1897.
35. Zupink: Berliner klinische Wochenschrift. No. 50. 1897.
36. de Martini: Centralblatt für Bacteriologie. Bd. XXI. 1897.
37. Spronck: Semaine médicale. 1897. No. 45. Abstract in Hygie-
    nische Rundschau. Jahr. VII, p. 1126.
38. Fraenkel: Hygienische Rundschau. Jahr. VII, p. 1126.
39. Muir and Ritchie: Manual of Bacteriology. P. 345.
40. Glücksmann: Zeitschr. für Hygiene. Bd. XXVI, p. 417.
41. Axenfeld: Berliner klinische Wochenschrift. No. 9. 1898.
42. Heinersdorff: Centralblatt für Bacteriologie. Bd. XXIII, p. 397.
    1898.
43. Kruse in Flügge: Die Mikro-organismen. Leipzig, 1896.
44. Schütz: Berliner klinische Wochenschrift. Jahr. XXXV, p. 297.
    1898.
45. Ehret: Münchener medicinische Wochenschrift. Jahr. 44.
46. Auckenthaler: Centralblatt für Bacteriologie. Bd. XXIII, p. 641.
    1898.
47. Peckham: The Journal of Experimental Medicine. Vol. II, No. 5.
    1897.

Publications

OF THE

University of Pennsylvania

Group I.—Annual Publications.

University Catalogue (published in December).

Fasciculi of the Departments of Philosophy (Graduate School), Law, Medicine, Dentistry and Veterinary Medicine; also Circulars of Information concerning courses offered in the College: No. 1 (School of Arts); No. 2 (Towne Scientific School); No. 3 (Courses for Teachers).

Report of the Provost (published in January).

Group II.—Serial Publications.

Series in Philology, Literature and Archæology.
Series in Philosophy.
Series in Political Economy and Public Law.*
Series in Botany.
Series in Zoölogy.
Series in Mathematics.
University Bulletin (monthly).

Group III.—Occasional Publications.

Reports of the Museums of Archæology and Paleontology.
Theses presented for the Degree of Doctor of Philosophy.

†Group IV.—Affiliated Publications.

Annals of the American Academy of Political and Social Science.
Americana Germanica (quarterly).
Bulletin of the Free Museum of Science and Art.
Translations and Reprints from the Original Sources of European History.

EXPLANATORY.

Group I consists of publications issued annually under the direct auspices of the Provost and Trustees.

The University Catalogue is a volume of about 500 pp. It contains detailed information concerning all departments, lists of officers and

---

* Beginning with New Series, No. 1.
† For exchange purposes only.

students, with addresses, etc. No charge is made for the Catalogue, but in all cases requests for a copy by mail must be accompanied by ten cents in stamps to cover postage.

The Fasciculus of each department contains information concerning that department *only;* while the three College Circulars of Information, covering respectively the School of Arts, the Towne Scientific School, and the Courses for Teachers, are in like manner restricted as to their contents. The Fasciculi and College Circulars are published separately after the University Catalogue, of which they are, to a large extent, reprints. Single copies are mailed free upon request.

The Report of the Provost, made by him annually to the Corporation, constitutes a general review of University activities during the year, and contains *inter alia* reports from the Treasurer and the several Deans. Single copies are mailed free upon request.

Group II consists of a number of serial publications in the several fields of literature, science and philology. They are issued in separate series at irregular intervals (for the most part), and represent the results of original research by, or under the direction of, members of the staff of instruction of the University. A complete list of these publications to date, *with prices attached*, is printed at length following. They are published under the editorial supervision of the University Publications Committee.

Group III consists of occasional publications, such as reports of the various University departments (where printed separately), and certain theses presented in partial fulfillment of the requirements for the degree Doctor of Philosophy.

Group IV consists of affiliated publications, issued as separate periodicals, not under the control of the University, but edited in part by officers of the University of Pennsylvania. Copies are obtainable from the University only through the medium of exchange (see Exchange Bureau, below).

EXCHANGE BUREAU.

The University of Pennsylvania desires to extend its system of exchanging publications with other similar institutions and learned societies, both at home and abroad.

For convenience in correspondence, the following statement is made:

To those educational institutions and learned societies which issue only annual catalogues, reports, or similar publications, the University of Pennsylvania offers in exchange all those of its own publications classed under **Group I** and **III**, or as many of them as may be specified.

To those educational institutions and learned societies publishing *also* results of original investigations, the University of Pennsylvania offers in exchange any one of its equivalent series in **Groups II** and **IV,** or as many of them as may be mutually agreed upon in order to maintain a proportionate ratio of exchange.

In establishing a system of exchanges with any other institution, the University of Pennsylvania binds itself to the following regulations:

All publications agreed upon to be forwarded from Philadelphia to address furnished, immediately upon issue, free of expense to our correspondent.

In return the University requests compliance with the following:

All publications to be forwarded to "Library of the University of Pennsylvania, Philadelphia, Pa.," marked "Exchange Bureau" in lower left-hand corner, immediately upon issue, free of expense to us.

Orders for single numbers, or sets of Serial Publications under **Group II,** and all correspondence relating to the publications of this University, should be addressed to

J. HARTLEY MERRICK, *Assistant Secretary,*
Station B, Philadelphia, Pa.

Philology, Literature, and Archæology

Volume I.

1. **Poetic and Verse Criticism of the Reign of Elizabeth.** By FELIX E. SCHELLING, Professor of English Literature. $1.00.
2. **A Fragment of the Babylonian "Dibarra" Epic.** By MORRIS JASTROW, JR., Professor of Arabic. 60 cents.
3. *a.* Πρός **with the Accusative.** *b.* **Note on a Passage in the Antigone.** By WILLIAM A. LAMBERTON, Professor of the Greek Language and Literature. 50 cents.
4. **The Gambling Games of the Chinese in America : Fán t'án and Pák kòp piú.** By STEWART CULIN, Secretary of the Museum of Archæology and Paleontology. 40 cents.

Volume II.

1. **Recent Archæological Explorations in the Valley of the Delaware River.** By CHARLES C. ABBOTT, Sometime Curator of the Museum of American Archæology. 75 cents.
2. **The Terrace at Persepolis.** By MORTON W. EASTON, Professor of English and Comparative Philology. 25 cents.
3. **The Life and Writings of George Gascoigne.** By FELIX E. SCHELLING, Professor of English Literature. $1.00.

Volume III.

1. **Assyriaca.** By HERMANN V. HILPRECHT, Professor of Assyrian and Comparative Semitic Philology and Curator of Babylonian Antiquities. $1.50.
2. **A Primer of Mayan Hieroglyphics.** By DANIEL G. BRINTON, Professor of American Archæology and Linguistics. $1.20.

5

Volume IV.

1. **The Rhymes of Gower's Confessio Amanti.** By MORTON W. EASTON, Professor of English and Comparative Philology. 60 cents.
2. **Social Changes in the Sixteenth Century as Reflected in Contemporary English Literature.** By EDWARD P. CHEYNEY, Assistant Professor of History. $1.00.
3. **The War of the Theatres.** By JOSIAH H. PENNIMAN, Instructor in English. $1.00.

Volume V. $2.00.

Two Plays of Miguel Sanchez (surnamed "El Divino"). By HUGO A. RENNERT, Professor of Romance Languages and Literatures.

Volume VI. $2.00.

a. **The Antiquity of Man in the Delaware Valley.**
b. **Exploration of an Indian Ossuary on the Choptank River, Dorchester Co., Md.** With a description of the crania discovered by E. D. Cope; and an examination of traces of disease in the bones, by Dr. R. H. Harte.
c. **Exploration of Aboriginal Shell Heaps on York River, Maine.** By HENRY C. MERCER, Curator of the Museum of American Archæology.

Philosophy ·

1. *** Sameness and Identity.** By GEORGE STUART FULLERTON.
2. *** On the Perception of Small Differences.** With special reference to the Extent, Force, and Time of Movement. By GEORGE STUART FULLERTON and JAMES MCKEEN CATTELL.

Political Economy and Public Law

†Volume I.

1. **The Wharton School Annals of Political Science.** March, 1885.
2. **The Anti-Rent Agitation in the State of New York.** 1839–1846. By EDWARD P. CHEYNEY.
3. **Ground Rents in Philadelphia.** By EDWARD P. ALLINSON and B. PENROSE.
4. **The Consumption of Wealth.** By SIMON N. PATTEN.

* Out of print.
† No copies available for exchange.

5. **Prison Statistics of the United States for 1888.** By ROLAND P. FALKNER.
6. **The Principles of Rational Taxation.** (Read at a meeting of the Association, November 21, 1889.) By SIMON N. PATTEN.
7. **The Federal Constitution of Germany.** With an historical introduction, translated by EDMUND J. JAMES.
8. **The Federal Constitution of Switzerland.** Translated by EDMUND J. JAMES.

Volume II.

9. **Our Sheep and the Tariff.** By WILLIAM DRAPER LEWIS.

Volume III.

10. **The German Bundesrath.** A Study in Comparative Constitutional Law. By JAMES HARVEY ROBINSON.
11. **The Theory of Dynamic Economics.** By SIMON N. PATTEN.

Volume IV.

12. **The Referendum in America.** A Discussion of Law-Making by Popular Vote. By ELLIS PAXSON OBERHOLTZER.

Volume V.

13. **Currency Reform.** By JOSEPH FRENCH JOHNSON. 25 cents.

CONTRIBUTIONS FROM

The Botanical Laboratory

Volume I—No. 1. $2.00.

(Plates I-XIII.)

1. **A Monstrous Specimen of Rudbeckia hirta, L.** By J. T. ROTHROCK, B.S., M.D.
2. **Contributions to the History of Dionæa Muscipula, Ellis.** By J. M. MACFARLANE, D.Sc.
3. **An Abnormal Development of the Inflorescence of Dionæa.** By JOHN W. HARSHBERGER, A.B., B.S.
4. **Mangrove Tannin.** By H. TRIMBLE, PH.M.
5. **Observations on Epigæa repens, L.** By W. P. WILSON, D.Sc.
6. **A Nascent Variety of Brunella vulgaris, L.** By J. T. ROTHROCK, B.S., M D.
7. **Preliminary Observations on the Movements of the Leaves of Melilotus alba, L., and other plants.** By W. P. WILSON, D.Sc., and J. M. GREENMAN.

*No copies available for exchange.

CONTRIBUTIONS FROM

The Zoölogical Laboratory.

Mathematics.